LEGO®

Holiday
IDEAS

Written by
Elizabeth Dowsett

Contents

Happy holly-days!

Holiday countdown

3, 2, 1—let's celebrate! Make the most of the run-up to the holidays with a LEGO® model that helps you count down to the big day.

Snowflake tree topper

This model has a tree at the top, but you can build anything you like

HOW TO USE

1 Build a multilevel structure with seven jumper plates that climb up to a summit.

2 Each day of the week before the holidays, move your minifigure up one spot.

3 When your minifigure reaches the top, it's time to celebrate!

Dripping snow and ice

Icicles clip into snow-colored brackets

Snowy animals

Why should snowmen have all the fun?
Make any creatures you like from LEGO
"snow!" Any collection of white bricks
and pieces you have can be turned
into snowy winter wildlife.

SNOWY OWL

Building snow
creatures is
a hoot!

★
TRY THIS

Build anything you
like from "snow." Try
building a snow pet, or
even a snow dinosaur.
There's "snow" limit
to what you
can create!

Colored pieces
make a festive
bow tie

Unicorn horn
is a blue transparent
lightsaber piece

Each ear is a
flat tooth plate

SNOWY UNICORN

Make claws
using bricks
with clips

Sleepy
eyes make
the unicorn
look like
he's resting

Front leg is a
curved slope piece

Showstopping snowflakes

Sprinkle some wintry magic in your home with these hanging snowflakes. They will bring your windows or tree to life, and they won't melt or drip on the floor.

Each arm is connected using hinge plates in the center of the ring

The snowflake's six arms are built using the same pieces

Transparent pieces add an icy sparkle

Use hinge plates to curve the sides

★ **TOP TIP**

Start in the middle with a rigid shape, such as a circle. Then build outward in a symmetrical pattern. Build up your snowflake one layer at a time.

Clip hinge plates onto plate with bars

Glowing
candles

Candles light up dark winter nights and are a part of many winter holiday traditions. Create a warm glow with these colorful candlesticks.

Use orange or yellow transparent pieces to make a flame

Radar dish makes a drip shield

Stack 2x2 round bricks to make the candle

Build a festive candlestick to display your candle

Ready, set, glow!

★ TRY THIS

Take out a few of the round bricks from your candle from time to time. This way, your candle will look like it's melting!

Circular baseplate holds the stand together

Welcoming wreaths

There's nothing like a colorful wreath hanging from your door to welcome guests! You can decorate your LEGO wreath however you choose.

Thread has two studs that clip between layers of plates

Overlapping circles make a stable ring

Decorate your wreath with your favorite colors

Green leaf piece

Brown corner plates go in opposite directions to the green corner plates

REAR VIEW

Red stud for berry

Green corner plate

Brown corner plate

Center of the bow is a brick with side studs

Decorated
gift boxes

Create a clever gift box for a friend's present. It's like wrapping paper they can keep! A gift box could even be a hiding place for your own wish list for Santa.

Tail pieces make a bow

Center the bow with a turntable piece

Tiles make the ribbon look smooth

Use bricks with clips to create a gift tag

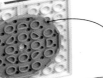

A circle plate holds the lid together

Tiles create a smooth look

All wrapped up

Build your box in the right shape and size for your gift to fit inside. This square box is just right for a small trinket such as a key chain.

Lid sits on top so it can be easily removed

I could use one of these for Mrs. Claus's present!

Red 2x2 round brick forms center of bow

The Elf
Express

Santa's elves are always busy up at the North Pole! They ride a mini monorail to zip from place to place over the snow-covered ground and keep things moving.

★ TOP TIP

Make the track first. Then build the monorail train from the center outward. Begin with a section that is the same width as the track.

Monorails have a single track

Turntable piece makes a snowy base

Use log bricks to make pylons

Flags tell elves which train they're catching

LEGO® Technic cylinder for a chimney

We're on track to deliver these gifts!

Cargo box holds gifts that need to be moved to Santa's sleigh

Attach wheels sideways so the train straddles the track

Green wedge attaches to the side of the train

Merry microfigures

Holiday decorations come in all shapes and sizes. These tiny microfigures are made up of simple pieces, but they create instantly recognizable winter characters to play with, share, or display.

Inverted slopes stick out for elvish ears

Does this hat make my ears look big?

ELF

★ **TOP TIP**

Keep all your characters to scale by using the same-size base for each microfigure. Then build upward. These are all built on 2x2 plates.

Create Santa's signature pom-pom with a white 1x1 plate

2x2 round brick makes the perfect hat

Stack white bricks to make long white hair

SANTA

SNOWMAN

MRS. CLAUS

Elf door

Invite elves into your home with a tiny door.
Decorate your door however you wish—
just make sure it's elf-sized. Place the door
against a wall and see what happens . . .

Tiles decorate the entryway

The back of a 2x2 turntable piece looks like a round window frame

Lamp to light the elves' way

I love to make an entrance!

Telescope pieces make fairy-tale-style hinges

Memory
match-up

Test your memory with this fun holiday card game. Make cards with these patterns or choose any festive images you like. Just remember to have fun!

★
TOP TIP

Make sure the bases are all the same size and color. When the cards are turned over, they must look identical.

Corner plate makes the top of an antler

Make sure each pair is identical

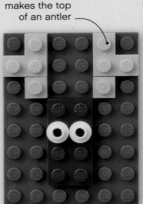

Why are there no elf cards?

Use small pieces to create simple, holiday-themed pictures

HOW TO PLAY

1 Make a deck of cards with an identical pair for each design.

2 Mix the cards and lay them face down on a flat surface.

3 Each person takes a turn to choose two cards. If they match, the player keeps them. If they don't, then they are turned back over.

4 The winner is the player with the most pairs at the end.

I can't remember!

A 6x8 plate makes a great base for the card

Center the trunk with a jumper plate

Studs next to smooth tiles give the look of dripping wax

Festive food market

Holiday markets are the place to find all your favorite festive food. Create your own bustling scene with LEGO pieces. You can almost smell the cinnamon spice!

TRY THIS

Add more carts and stands to make a bigger market. What kinds of food and drinks will you serve up?

Croissant sign

Minifigures can gather around this barrel for a hot chocolate

Angled tooth plates make a frilly canopy

Transparent windshield makes a pastry case

Leg support connects to a clip hinge so it can be folded away for travel

Connect wheels to make your cart mobile

PASTRY CART

Crystal ball piece

Transparent cone

2x2 round plate

Brick with studs on four sides holds lamppost together

Sugar, spice, and everything nice

Wooden stands are inspired by Alpine huts. Market vendors sell food, drinks, handmade crafts, and gifts.

Twinkling lights made from colored studs

Decorated lamppost

Ketchup bottle

ALPINE SAUSAGE STAND

Evergreen decorations

German-style sausages are often sold at holiday markets

Ho-ho-homemade games

Have fun making these board games to play with family and friends. Use your LEGO pieces to make a festive-themed tic-tac-toe board and pieces or a slippery ice game.

HOW TO PLAY

1 Build a board with nine squares. Build five game pieces of one design for one player and five of a second design for the other player.

2 Two players take turns to place a game piece on the board.

3 The winner is the first person to place three of their pieces in a row. Rows can be vertical, horizontal, or diagonal.

Tic-tac mistle-toe!

Challenge your friends to a game of tic-tac-toe. Swap traditional Xs and Os for festive pieces like these.

Three in a row wins!

Turkey dinner piece

Gift game piece

Each space on the board is a jumper plate

HOW TO PLAY

1 Two players take turns to roll the die and move their elf around the course.

2 If a player lands on the same square as their opponent, their opponent falls into the water and has to go back to the start.

3 If a player lands on a piece of ice, they slip into the water and have to go back to the start.

4 The winner is the first player to reach Santa at the North Pole.

Go with the flow!

Two elves are stranded in an ice drift! Can you help them hop their way over the ice floes to join Santa?

★ TRY THIS

You could use a regular die, or why not build one from LEGO bricks? Use single studs for the dots on each side.

The "North Pole" marks the finish line

I gotta catch up to that elf on the ice shelf!

Elf game piece

Starting square

Start with blue baseplate for the water

Landing on icy transparent tile means players have to start again

Gingerbread
train

Whip up a treat for your eyes with this delicious-looking gingerbread train. What a delectable decoration!

Alternate layers of brown and white to give the look of gingerbread and icing

Now that's a train I can get on board with!

Transparent round plates make good lights

The train is built from the bottom up, starting from this baseplate

Wheels are two round plates stacked on top of each other

UNDERSIDE VIEW

Wheels build onto this gray plate attached sideways

Yellow transparent bricks create a window that looks like it's made from hard candies

⭐
SPECIAL BRICK

1x1 transparent round plates are perfect for making the gingerbread look like it's decorated with candy.

White tooth pieces look like piped icing

Winter
warmer

Yodelayheehoo! A cold day calls for a hot treat. Build a snack truck to keep your minifigures warm on the ski slopes. Hot chocolate, anyone?

Main roof easily lifts off for access to the inside

1x1 plates with clips attach to the string and hold transparent pieces in place

Tile for a side-view mirror

Add festive touches, such as a wreath

TRY THIS

Expand your scene! Add skiers, slopes, a ski lift, food vendors, and even some places to relax after your minifigures have hit the slopes.

Barrel makes an outdoor table

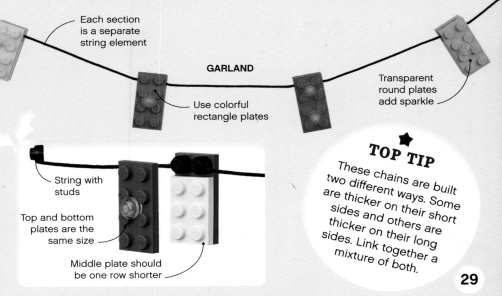

PAPER CHAIN

Single plate

Connect as many rectangles as you like to reach your desired length

Stack of three plates

Festive
flourishes

String up some holiday cheer using the most basic pieces. All you need for a bright paper chain is simple rectangles. With LEGO strings, you can make a garland.

Each section is a separate string element

GARLAND

Use colorful rectangle plates

Transparent round plates add sparkle

String with studs

Top and bottom plates are the same size

Middle plate should be one row shorter

★ TOP TIP

These chains are built two different ways. Some are thicker on their short sides and others are thicker on their long sides. Link together a mixture of both.

Meet
Santa

When Santa comes to town, everyone lines up to meet him. Build Santa a comfy chair so your minifigures can have their picture taken with him. Say "cheese!"

★
TRY THIS

Expand your scene! You could build the toy section of the department store. And why not park Santa's sleigh and reindeer nearby?

Chains on round bricks guide the line

What did you ask Santa to bring you this year?

Use plates to create a red carpet for visitors

Make a simple wreath with a round plate

Attach the hinge brick to the plate

Hinge brick and plate create the angle of the roof

Log brick

Candy stripe roof trimming made with 1x1 round bricks

A big white plate makes the roof look like it's covered in snow

Elf photographer

Stocking-shaped gift box

31

Party animals

The holidays are a time for family—including pets! Build yourself some cute little LEGO pets and give each animal some festive finishing touches.

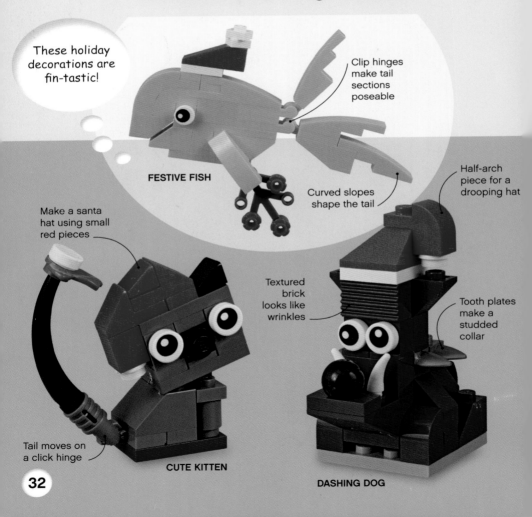

These holiday decorations are fin-tastic!

Clip hinges make tail sections poseable

FESTIVE FISH

Half-arch piece for a drooping hat

Curved slopes shape the tail

Make a santa hat using small red pieces

Textured brick looks like wrinkles

Tooth plates make a studded collar

Tail moves on a click hinge

CUTE KITTEN

DASHING DOG

Terrific tree

Branch out with your decorations this year. It takes years to grow a good fir tree, but you can build yourself a LEGO tree in minutes. This one will brighten up any little space, such as a window sill.

Top your tree with a special piece, such as a snowflake

Build flat sections in increasing sizes

Turntable piece sits between all but the top two sections

Stack tree layers from the base upward

Make ornaments from round plates

Radar dish makes a sturdy base

1x2 textured bricks give the appearance of a real chimney

This is a tight squeeze. Maybe I shouldn't have eaten that last cookie . . .

Green circular plates are surrounded by leafy pieces

Building with round bricks creates textured columns

Leave stockings by the hearth

Clip looks like a ribbon bow

FRONT VIEW

Chimney surprise!

Now you see him . . . now you don't! Dive into this build headfirst and amuse your friends with a Santa who appears and disappears up the chimney.

Use the handle to slide Santa up and down the chimney

Jumper plate makes a beard with a surprised expression

Two 2x2 round bricks form the handle

SANTA

Create Santa's jolly tummy with a round plate

Slope bricks shape the fireplace

REAR VIEW

Build your fireplace onto a baseplate

All the trimmings

Spruce up your tree with these festive hanging ornaments. Start by building flat shapes of any size and you'll soon get the hang of making decorations!

String allows the decoration to hang

Plate with clip

TREE

Secure the string element with a slide plate

String with studs connects to the back

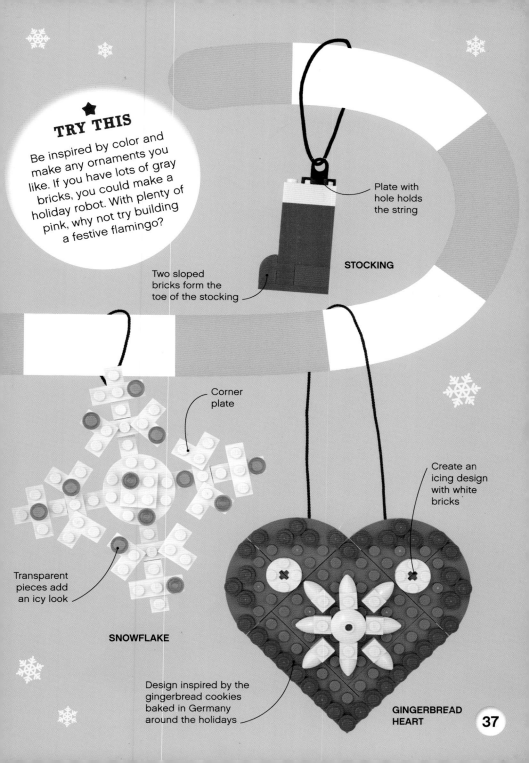

★ TRY THIS

Be inspired by color and make any ornaments you like. If you have lots of gray bricks, you could make a holiday robot. With plenty of pink, why not try building a festive flamingo?

Plate with hole holds the string

STOCKING

Two sloped bricks form the toe of the stocking

Corner plate

Create an icing design with white bricks

Transparent pieces add an icy look

SNOWFLAKE

Design inspired by the gingerbread cookies baked in Germany around the holidays

GINGERBREAD HEART

Spot the difference

How observant are your friends and family? Find out with a holiday-themed 3D puzzle—like this rooftop scene with Santa heading down the chimney.

Bright full moon

Make stars with transparent studs

SCENE 1

Stack textured bricks to make a chimney

Do you see what I see?

When building differences for your friends to find, think about the color, size, and position of bricks, missing objects, and swaps. There are seven differences to spot in these models, but you can build in as many differences as you like.

HOW TO PLAY

1 Build two models that are identical apart from a few small details.

2 Challenge a friend to spot all the differences. Remember to tell them how many differences to look for.

3 Play again and again by changing what the differences are.

Yellow and green transparent studs for stars in the sky

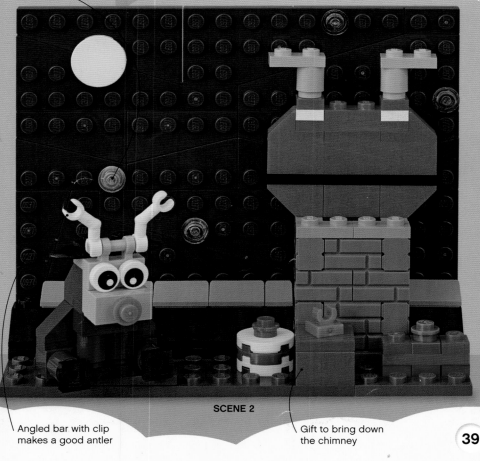

SCENE 2

Angled bar with clip makes a good antler

Gift to bring down the chimney

39

Festive build
challenge

Practice your building skills and test your creativity with this fun LEGO challenge. You can play this game all year round, but during the holidays, give it a festive twist!

HOW TO PLAY

1 Ask each player to write down five building ideas on pieces of paper and fold them up.

2 Place the ideas in a bag. Each player chooses one at random.

3 Set a timer for five minutes. Each player must build the idea on their paper before time is up.

4 When time is up, everyone must guess what each player has built.

5 If someone correctly guesses a build, the guesser and the builder each get two points.

6 The winner is the person with the most points at the end of the game.

Ribbons fit onto hose connector

Domes look like bells

HOLIDAY BELLS

Set out a random pile of bricks to make your builds

We're off to a flying start with these creations!

FESTIVE ROBIN

Create an iconic red breast with a round tile

Use green plates to make holly leaves

White bricks look like the custard

TRADITIONAL ENGLISH CHRISTMAS PUDDING

Goblet for a festive toast

Edge plate for a tail

Build and be merry!

FESTIVE LOBSTER

Tiles give the notes a smooth look

★ TRY THIS

To mix it up, give everyone the same idea to build. Then see how different or similar each model is. There's always more than one way to build something.

HOLIDAY MUSIC NOTES

Santa's workshop

With gifts to make for girls and boys around the world, Santa's elves are always busy! Build a tiny workshop where your minifigure elves can make lots of tiny toys.

Tools held with clips

Headlight brick

Toy soldier

DOLL HOUSE

Jumper plates look like door handles

TOOL BENCH

Add an ornate finish with a scroll brick

Make paint and glue pots with 1x1 round bricks

WORKSTATION

Elves' shelves

The elves need lots of materials and tools to make toys. Build some shelves to keep the workshop neat and tidy.

Gold plant-shoot

This piece was used in a LEGO® NINJAGO® crown and comes in lots of colors

Join two fence pieces with tiles to make a ladder

TOY SHELVES

Fill the shelves with brick-built toys and tools

Plate with clip for ears

Tail attaches to a headlight brick

This horse really rocks!

Macaroni tile for a rocker

ROCKING HORSE

The Wrap-O-Matic

After all the toys have been made, it's time to wrap them! Why not build the elves a special wrapping machine to speed things up? Ready, set, wrap!

Okay, everyone. That's a wrap!

★ **TOP TIP**

Build gifts with just two plates plus a jumper plate on top. If you sandwich together two colors, the middle layer looks like a ribbon. Attach a bow to finish the gift.

Stack plates and tiles to make wrapped gifts

Fill a crate with wrapped presents

Flower stud makes a frilly bow

Steam escapes
from the engine

Rolls of
ribbon

Scissors
ready for
cutting

Yellow light
means the
machine is
turned on

Wrapped gift,
ready for
Santa's sleigh

Present
about to be
wrapped

Grooved tiles look
like a conveyor belt

★
SPECIAL BRICK

LEGO gears make things move like clockwork, but you don't have to use them for movement. Adding them like this gives a machine a technical look.

45

Attach the candy cane using a headlight brick

Brick with side stud

Add a big statement piece, like a giant candy cane

Hang "lights" on a string with bars

Icicle

Snowdrift piled up by the fence

FRONT VIEW

Holiday
home makeover

Deck the halls—and LEGO houses!
With some wintry touches and sparkly
flourishes, you can transform any LEGO
building into a festive one.

Add white
bricks to cover
your house in
snow

★
TRY THIS

You can build a simple
house and decorate it.
Or why not use any
existing LEGO houses or
buildings you have and
dress them up for
the holidays?

Each roof panel
connects with
a click hinge

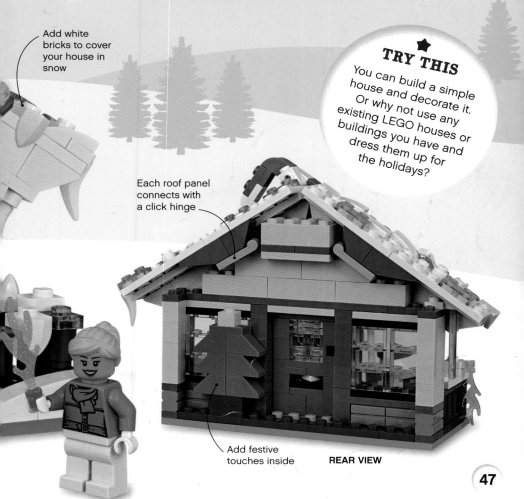

Add festive
touches inside

REAR VIEW

Festive
family dinner

Festive food is a popular part of celebrations all around the world. Turkey or tofu, cookies or panettone—whatever your traditions, build a tempting feast for minifigures of all ages to share!

I can't wait to gobble up that turkey!

Brick with side studs for turkey's body

Round jumper plate

DINING TABLE AND CHAIRS

Make as many dining chairs as you need

Light up the room with flame pieces

Cone makes a great candle

★
TRY THIS
There are plenty of LEGO food elements available, but you can also use standard bricks to make your family's favorite foods. What will you cook?

Diamond jewel element makes an elegant bottle stopper

Small radar dishes look like dessert plates

DESSERT TABLE

DRINKS TABLE

Chair piece often comes in vehicle sets

Better get more cookies—Uncle Larry brought turkey surprise!

Log brick

Use a small plate for a base

Winter
wonderland

Bring the outside inside with a snowy nature scene. Try building your landscape in full color first to think about the shapes you need to recreate the natural world.

Stack slopes in different directions to make trees

Round plates separate each layer of the evergreen tree

Make antlers by stacking small plates with clips at different angles

Rocky outcrop

Brown plates for earth underneath grass

Patches of melting ice

Flower stem pieces add texture

Let it snow!

Now turn your colorful woodland into a frozen, snow-covered scene. Using just one color makes a striking build.

All white bricks give the appearance of a snow-covered tree

Add wintry elements to your frosty scene, such as a snowy bird

Snowdrift

Create the look of fresh-fallen snow with smooth tiles

Horn pieces look like snow-covered shoots

Jumper plate holds the minifigure in place

Snow bunny knows how cool this wintry scene is!

★ **TRY THIS**

Winter holidays are celebrated around the world in many climates. You could create a holiday scene in the rainforest, in the desert, or on the beach.

Cool crackers

Decorate your table with a LEGO cracker for each guest. These special party favors are a holiday tradition in some countries. Pull the crackers apart to reveal little gifts inside!

★ SPECIAL BRICK

The Golden Viper mini-snake element from the LEGO NINJAGO theme makes a perfect swirly ribbon.

Red ball looks like a holly berry

Ribbon tail is held in place with a clip

One red 2x8 brick connects the "twist" to the main cracker

CHRISTMAS CRACKER

Each half of the cracker is built on one 6x4 plate

This is a cracking idea for a model!

Pull out all the stops

LEGO® Technic pin-and-hole connections make sturdy crackers that pull apart easily.

LEGO Technic half pin

INSIDE VIEW

1x1 brick with hole

Slopes form end of cracker

TRY THIS

Fill your cracker with mini-builds of traditional gifts, such as a golden crown. Once the cracker is opened, use the trinkets for a treasure hunt.

PENGUIN

Use colorful tiles to make a gift

GIFT

A round plate makes tiny penguin feet

CROWN

Gold pyramid makes an iconic crown shape

Roll up your letter to fit it in the holder

Decorate your pens for the holidays

Store your pens here

Letters for Santa

What's on your holiday wish list? More importantly, where *is* your wish list? Before you mail it to the North Pole, keep your list safe with a special letter holder, like this one. You can store your pens inside it, too!

TOP TIP

The green section is built with bricks that are two studs wide. The red cuff is built larger so it overhangs the top like folded-over fabric.

Inverted slopes shape the heel

Silly selfie props

'Tis the season to be jolly and take lots of fun photos. Make some playful props for you and your friends. Then strike a festive pose!

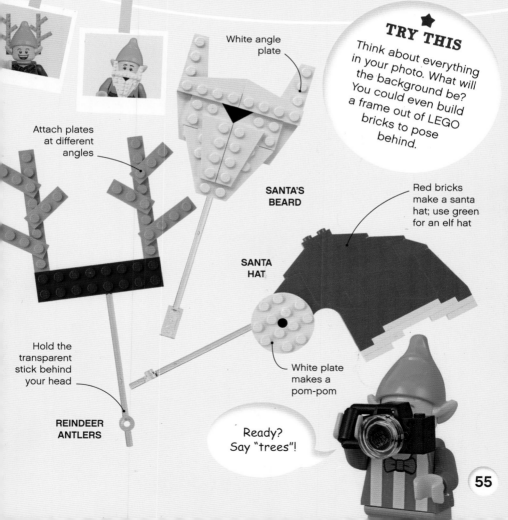

White angle plate

⭐ **TRY THIS**

Think about everything in your photo. What will the background be? You could even build a frame out of LEGO bricks to pose behind.

Attach plates at different angles

SANTA'S BEARD

Red bricks make a santa hat; use green for an elf hat

SANTA HAT

Hold the transparent stick behind your head

White plate makes a pom-pom

REINDEER ANTLERS

Ready? Say "trees"!

Elf obstacle course

These speedy elves are competing to find out who the fastest elf at the North Pole is. Will you build these challenges or invent some of your own?

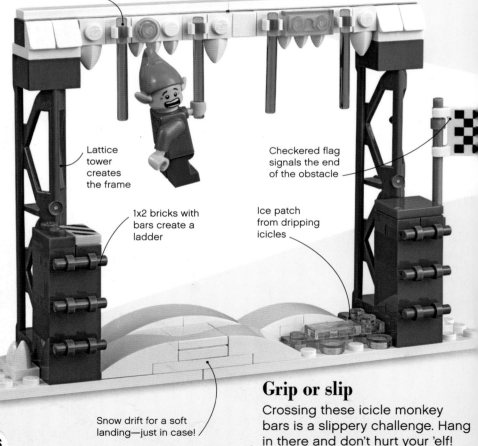

Use transparent blue bars for icicles

Lattice tower creates the frame

1x2 bricks with bars create a ladder

Checkered flag signals the end of the obstacle

Ice patch from dripping icicles

Snow drift for a soft landing—just in case!

Grip or slip

Crossing these icicle monkey bars is a slippery challenge. Hang in there and don't hurt your 'elf!

There's nothing like a little "elfy" competition!

Delivery dash

Wind the gear to rotate the conveyor belt. The elf must hop over the ridges to get the gift to the finish line.

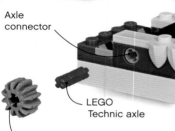

Axle connector

LEGO Technic axle

LEGO Technic gear

Rubber track winds around two cylinders to make a moving conveyor belt

Red tiles create a finish line

Layered leaf pieces make a sturdy wreath

Holly hoop

It's the final challenge! Elves must run up the stairs and leap through the wreath without touching its sides. Who will win?

Sideways-built stairs capped with smooth tiles

1x1 red flower for holly berries

Hidden
elf

Small and stealthy, elves are very good at hiding! Challenge your friends to find an elf minifigure in this fun hide-and-seek game.

Build your elf cups with enough height to fit a minifigure inside

Decorate your cups however you like

Wedge pieces around a 1x1 brick make a good bow

A baseplate covers the top

Each cup should be hollow

HOW TO PLAY

1 Build three identical cups, each with four walls and a plate on top. Leave the bottom open.

2 Hide an elf minifigure under one of the cups.

3 Place the cups in front of a friend.

4 Shuffle the cups around in front of your friend.

5 Ask them which cup holds the elf.

A flurry of snowmen

Just like snowflakes, no two snowmen are the same. Challenge yourself to see how many different snowmen you can build. Best of all, these LEGO snowmen won't melt away!

★ **SPECIAL BRICK**

This white plate with black markings makes a great snowman tooth. If you don't have this piece, you can use a whole black plate instead.

LEGO Technic pin in a brick with a hole secures nose

Carrot piece

Icing swirl piece for pom-pom

Small slope makes the scarf 3D

Stack rectangular shapes for a square-shaped torso

Black wrench pieces for arms

Minifigure top hat

White dome is a stable base

Use round plates for a cylindrical body

Building is "snow" much fun!

Santa's sleigh

Get your holiday celebrations off to a flying start. Use your LEGO bricks to create the most magical vehicle of all time—Santa's sleigh! Of course, you'll also need a magical reindeer or two to pull Santa and his sleigh through the sky.

It's the most won-deer-ful time of the year!

★
SPECIAL BRICK

Hold the reindeer's carrot treat in place with a long, flexible hose piece, which has connectors on each end.

Pile up loose bricks to make gifts

Ho, ho, ho and away we go!

T-bar handle controls the carrot lead

Sleigh built in Santa's signature red and white

Hitch your reindeer to the sleigh with a gold chain piece

Cookie holder in case Santa gets hungry

Joystick

1x2 plate with bar

1x3 curved slope piece

Penguins
on parade

Think of winter, and you think of snow. Think of snow, and you think of penguins. Grab all your black, white, and orange bricks and see how big a parade of penguins you can make.

A dish makes a good tummy for a larger penguin

Black tooth pieces can be used as wings

Use round plates on their sides for sleepy eyes

⭐ **SPECIAL BRICK**

Penguins need big, sturdy feet to waddle on. Flippers from diver minifigures are ideal.

White tooth plates for penguin bellies

Ready for your photo, Mr. Penguin? Say "freeze"!

Slopes or tooth plates make perfect beaks

Candy cane decorations

These cute candy canes make the perfect hanging decoration. Their bright, seasonal stripes are a treat for the eyes!

For bright stripes, alternate strong colors

Hang the cane on your tree by its hooked top

A simple stack forms the base of the candy cane

Make some smaller canes from single stud bricks

Decorate your candy cane with festive trimmings

Swedish Dala
horse

The Dala horse is a carved and painted wooden horse from Sweden. The lively colors brighten up a wintry day, making it an eye-catching holiday decoration.

1x1 tooth plate for the mane

Mane and stirrup pieces clip onto bricks with side studs

Curved half-arch piece

Make a stirrup with a 1x2 tile

Stack rectangular bricks for the legs

TOP TIP

The most famous Dala horse is bright red with blue and white decoration. They come in all colors and patterns, so make yours however you like.

Sleeves are attached with angle plates

Sausage piece arms held in place with plates with clips

Transparent snowflakes

Cozy holiday sweaters

Festive sweaters are in fashion—and now you can build them from LEGO bricks, too! There's no limit to the designs you can create in LEGO "knitwear."

Hang sweaters on your tree with string through a plate with a hole

Rectangular sleeves are connected behind the body of the sweater

★ SPECIAL BRICK

Adding a curved plate with a hole to the back of the sweater means you can thread string through it to make a hanging decoration.

Yellow jumper plate for the belt buckle

Santa band

It's beginning to sound a lot like the holidays with this musical minifigure lineup. Santa rocks out with some talented elves to back him up while the Gingerbread Man plays piano.

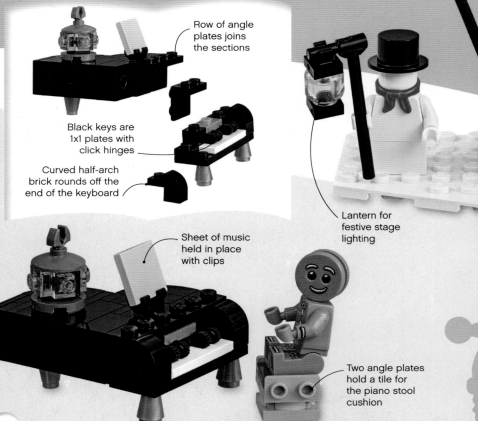

Row of angle plates joins the sections

Black keys are 1x1 plates with click hinges

Curved half-arch brick rounds off the end of the keyboard

Lantern for festive stage lighting

Sheet of music held in place with clips

Two angle plates hold a tile for the piano stool cushion

Use a leaf piece and white round plates to make mistletoe

Pole attaches to click-hinge cylinder

Small radar dish looks like a speaker

Build onto a row of bricks to give your stage some lift

Raised stage holds baseplates together

Half-arch piece
for palm frond

Corner tile with
an angled corner

Cross plate

Brick with
studs on all
sides

Cross axle
connects
the trunk

LEGO Technic
angle element

Round brick
with hole

Santa is
ready to surf

TRY THIS

What other vacations could Santa take? You could build him on safari, on a cruise, visiting famous landmarks, or skiing—if that's not too much like home!

Transparent blue
tiles for the surf

Santa's
beach getaway

By the time Santa's delivered the final gift,
he's in need of a nice, long rest.
Build him a sunny vacation
spot to get away from it all!

I'm not sure
this umbrella
is working . . .

White tiles on the ground
make the snowman look
like he's melting

Shovel for
building
sandcastles

Surf's up,
"sandy" Claus!

Tan plates make
a sandy beach

Sandcastle

Penguin
Random
House

Senior Editor Tori Kosara
Editor Rosie Peet
Senior Designer Lauren Adams
Designer Elena Jarmoskaite
Senior Pre-production Producer
Jennifer Murray
Senior Producer Lloyd Robertson
Managing Editor Paula Regan
Managing Art Editor Jo Connor
Publisher Julie Ferris
Art Director Lisa Lanzarini
Publishing Director Simon Beecroft
Model Photography Gary Ombler

Models created by
Barney Main and Rod Gillies

Dorling Kindersley would like to thank Randi
Sørensen, Heidi K. Jensen, Paul Hansford, Martin
Leighton Lindhardt, Nina Koopmann, Charlotte
Neidhardt, and Anette Steensgaard at the LEGO
Group; Kayla Dugger at DK for proofreading;
Sam Bartlett, Stefan Georgiou, James McKeag,
and Lisa Robb at DK for design assistance;
Victoria Taylor for editorial help; and model
builders Stuart Crawshaw,
Alice Finch, and Naomi Farr.

First American Edition, 2019
Published in the United States by DK Publishing
1450 Broadway, Suite 801, New York,
New York 10018

A WORLD OF IDEAS:
SEE ALL THERE IS TO KNOW

www.dk.com
www.LEGO.com

Meet the builders

Barney Main

Barney owns more than
40,000 LEGO® bricks!
He really enjoyed
building the holiday
sweaters. His top tip for
building holiday-themed
models is to use lots of
red, green, gold, and white.

Rod Gillies

LEGO fan Rod
particularly enjoyed
building the "Elf
obstacle course".
His tip for building
holiday-themed
models is to add snow,
ice, and anything sparkly.